KU-021-825
MARYBANK P. SCHOOL
No.

Animal Young

Mammals

Rod Theodorou

First published in Great Britain by
Heinemann Library,
Halley Court, Jordan Hill, Oxford OX2 8EJ
a division of Reed Educational and Professional
Publishing Ltd.
Heinemann is a registered trademark of Reed
Educational & Professional Publishing Ltd.

OXFORD MELBOURNE AUCKLAND
JOHANNESBURG BLANTYRE GABORONE
IBADAN PORTSMOUTH (NH) USA CHICAGO

Designed byCelia Floyd
Illustrations by Alan Fraser
Printed in Hong Kong/China

03 02 01 00 99
10 9 8 7 6 5 4 3 2 1

ISBN 0 431 03070 7

British Library Cataloguing in Publication Data

Theodorou, Rod
Mammals. – (Animal young)
1. Mammals – Infancy – Juvenile literature
I. Title
599.1'39

Acknowledgements
The Publishers would like to thank the following for permission to reproduce photographs:

BBC: Andrew Cooper p. 10, John Cancalosi p. 15, Anup Shah p. 17, Thomas D Mangelsen p. 23, Carl Englander p. 24; Bruce Coleman: p. 21, Mark Carwardine p. 5, Jane Burton p. 6, Rod Williams p. 13; Frank Lane: Silvestris p. 11; NHPA: A.N.T. p. 7, E A Janes p.12, Gerard Lacz p. 22, Michael Leach p. 26; OSF: Richard Kolar p. 8, Martyn Colbeck p. 16, Owen Newman p. 19, Mike Birkhead p. 20, Mark Deeble & Victoria Stone p. 25; Tony Stone: Renee Lynn p. 9, Art Wolfe p. 14, Norbert Wu p. 18.

Cover photograph reproduced with permission of Oxford Scientific Films/Konrad Wothe

Every effort has been made to contact copyright holders of any material reproduced in this book. Any omissions will be rectified in subsequent printings if notice is given to the Publisher.

Any words appearing in the text in bold, **like this**, are explained in the Glossary.

Contents

Introduction

There are many different kinds of animals. All animals have babies. They look after their babies in different ways.

These are the six main animal groups.

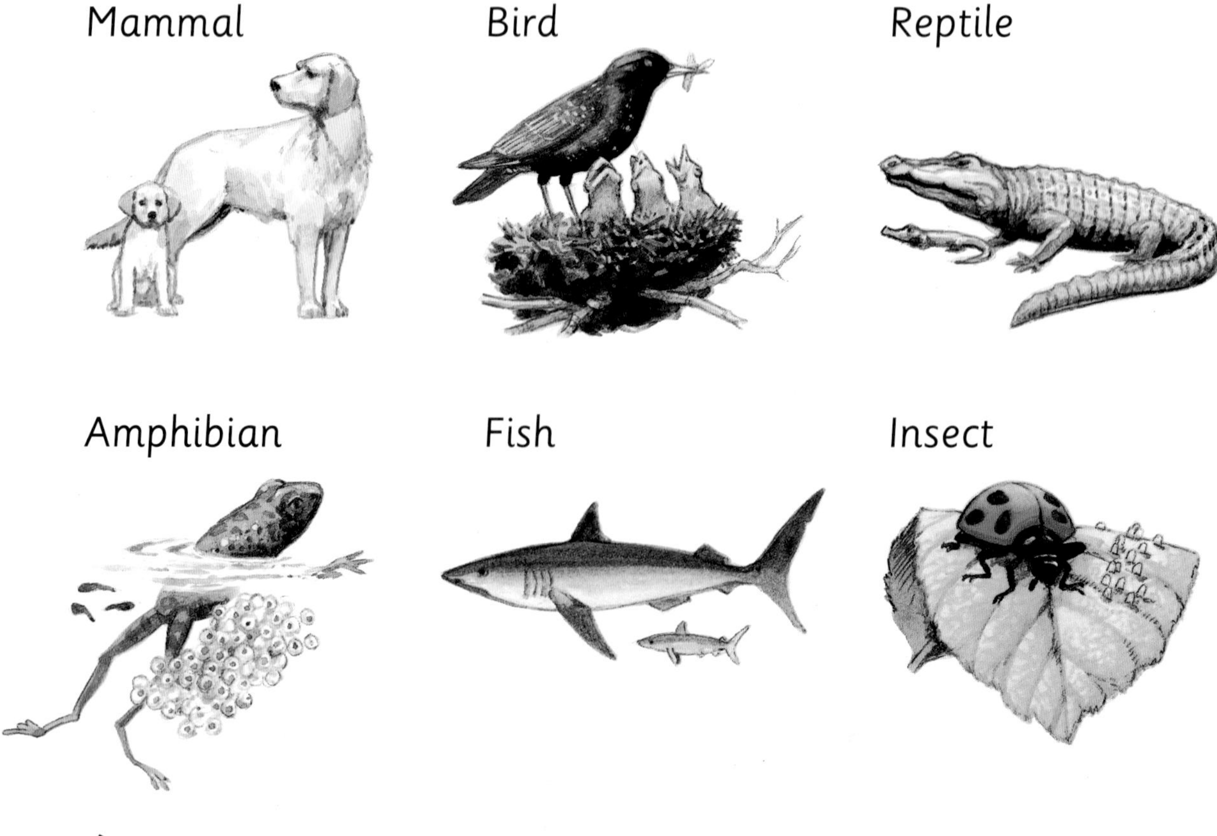

This book is about mammals. Mammals live all over the world. Some mammals are tiny. Some mammals are huge.

Adult blue whales are the biggest mammals in the world.

What is a mammal?

All mammals:

- breathe air
- feed their young milk from **teats** on the mother's body
- have hair on their bodies.

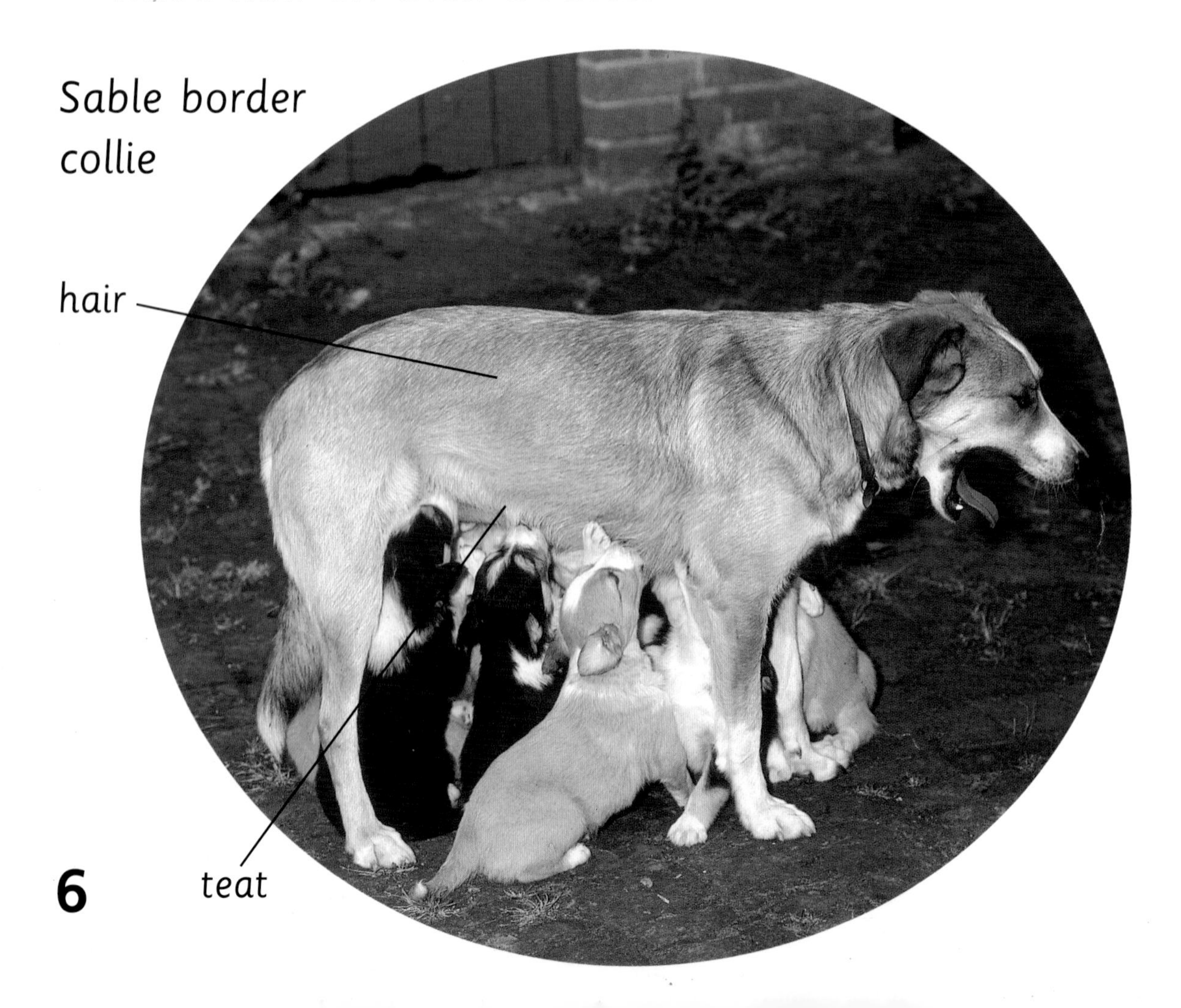

Sable border collie

Most mammals:

- grow babies inside the mother's body
- give birth to live babies
- live on land and walk on four legs.

Platypus babies ***hatch*** *from eggs, but they are mammals too!*

Birth

Some mammal mothers have one baby at a time. Others have lots of babies at once. Most mammals look for a safe place to have their babies.

Mice can have more than 15 babies in one ***litter****.*

Some mammal babies are born blind. Their parents look after them until their eyes open. Other mammal babies can see and run with their parents very soon after they are born.

The mother orang-utan carries her baby on her back until it can look after itself.

Looking after baby

Most mammals take care of their babies. Small mammals build burrows or nests for them to live in. They bring them food until they are old enough to find their own.

This mother fox has dug a burrow in the ground to keep her babies safe and warm.

Some mammals have to keep moving to find food or keep out of danger. Some carry their babies by the scruff of the neck. Other babies hold onto their mothers.

This baby bat holds onto its mother even when she is flying.

Feeding

Mother mammals feed their babies milk from their **teats**. The milk is very rich and helps the babies grow quickly.

The teats nearest this mother pig's head give the best milk.

Mammals need lots of food for **energy** and warmth. As the babies get older, their mothers **wean** them off milk. The babies have to start eating solid food.

These snow leopard babies are waiting for their mother to bring them meat.

Moving about

Some mammal babies can move around soon after they are born. They learn to walk or run so they can escape from **predators** and follow their parents.

Baby giraffes have to stand up and run soon after they are born.

Other mammal babies cannot walk or run when they are born. They stay close to their mother while they grow bigger and stronger.

Joeys leave their mother's **pouch** when they are about nine months old, but they jump back in if they are scared or hungry.

Family life

Many mammals live in large family groups, called **herds**. They help each other look after the young and **protect** them from danger.

When the herd go to look for food, some female elephants stay behind to 'babysit' all the young elephants.

Mothers in a herd always know which are their babies by their look, smell, and call. They spend a lot of time **grooming** and cleaning the babies.

Chimpanzee mothers pick dirt and insects out of their babies' hair.

Brothers and sisters

When they are young, most mammals spend a lot of time with their **siblings**, especially if they live in a **herd** or family group.

Young dolphins swim with their siblings. They may even fight and 'bully' young dolphins from other families or groups.

Brothers and sisters sometimes have play fights. This helps them grow stronger. It is also good practice for fighting **enemies** or for hunting **prey**.

This game will help these lion cubs learn to hunt and catch food.

Learning from parents

Some mammal babies have very good **instincts**. They do not have to learn everything from their parents. They know how to find food and look out for danger.

A young mouse knows how to find food and stay hidden from ***predators****.*

Other mammals have to learn some things from their parents. Some remember how their mothers looked after them, so they can look after their own babies.

Whale calves can swim as soon as they are born, but their mother has to push them to the surface to teach them to breathe air.

Learning to feed

Many mammals teach their babies to hunt or find food. Hunting mammals sometimes bring small or **injured** animals back to their young so they can learn how to kill.

This tiger mother teaches her cubs how to hunt other animals.

Some baby mammals spend a lot of time watching how their mother finds food. They will remember what plants are good to eat or what animals are easy to catch.

These bear cubs watch carefully to see how their mother catches fish.

Danger

Most mammals try to **protect** their young from danger. They may call to their young if a hunter is near. The young stand near their parents or run to a safe place.

When musk oxen are attacked by wolves, the big adult oxen stand in a circle with their young safe in the middle.

Some mammals are camouflaged. This means the colour of their skin or hair makes them hard for **predators** to see.

Cheetah cubs have long hair that makes them hard to see in tall grass.

Leaving home

Some mammal babies take years to grow up. Others grow up very quickly. Once they have grown up all mammals **mate** to have babies of their own.

Voles are ready to have their own babies when they are only 15 days old!

When they are old enough, some mammals leave their mother and find a new place to live. Other mammals stay with their family or their **herd** all their lives.

Humans are mammals. We spend more time living with our parents than any other mammal.

Mammals and other animals

		Fish
What they look like:	Bones inside body	all
	Number of legs	none
	Hair on body	none
	Scaly skin	most
	Wings	none
	Feathers	none
Where they live:	Lives on land	none
	Lives in water	all
How they are born:	Grows babies inside body	some
	Lays eggs	most
How they feed young:	Feeds baby milk	none
	Bring babies food	none

	Amphibians	Insects	Reptiles	Birds	Mammals
	all	none	all	all	all
	4 or none	6	4 or none	2	2 or 4
	none	all	none	none	all
	none	none	all	none	few
	none	most	none	all	some
	none	none	none	all	none
	most	most	most	all	most
	some	some	some	none	some
	few	some	some	none	most
	most	most	most	all	few
	none	none	none	none	all
	none	none	none	most	most

Glossary

enemy an animal that will kill another animal for food or for its home

energy to be able and strong enough to run around and play or hunt

grooming to keep healthy by cleaning and brushing fur or hair

hatch to be born from an egg

herd large group of animals of one kind that live together

injured hurt

instinct to be able to do something without being told how to

joey a baby kangaroo

litter group of animals that are born at the same time and have the same mother

mate when a male and a female animal come together to make babies

pouch pocket of skin on the stomach of some animals in which their babies grow

predator an animal that hunts and kills other animals for food

prey an animal that is hunted by another for food

protect to look after

sibling brother or sister

teat part of a female animal's body which the babies suck to get milk

wean when a baby animal stops feeding on its mother's milk and eats other food

Further reading

Big and Small, Rod Theodorou and Carole Telford, *Animal Opposites*, Heinemann Library, 1996

Bear, Claire Robinson, *Really Wild*, Heinemann Library, 1997.

Dolphin, Claire Robinson, *Really Wild*, Heinemann Library, 1999.

Elephant, Claire Robinson, *Really Wild*, Heinemann Library, 1997.

Fast and Slow, Rod Theodorou and Carole Telford, *Animal Opposites*, Heinemann Library, 1996

Heavy and Light, Rod Theodorou and Carole Telford, *Animal Opposites*, Heinemann Library, 1996.

Lion and Tiger, Rod Theodorou and Carole Telford, *Spot the Difference*, Heinemann Library, 1996.

Polar Bear and Grizzly Bear, Rod Theodorou and Carole Telford, *Spot the Difference*, Heinemann Library, 1996.

Prickly and Smooth, Rod Theodorou and Carole Telford, *Animal Opposites*, Heinemann Library, 1996

Short and Tall, Rod Theodorou and Carole Telford, *Animal Opposites*, Heinemann Library, 1996

Whale, Claire Robinson, *Really Wild*, Heinemann Library, 1999.

Index